Elementary School Library Dept.
Brandywine Public Schools
Niles, Michigan

Insects

Insects

BY ALICE FIELDS
Illustrated by David Hurrell

An Easy-Read Fact Book
FRANKLIN WATTS
New York/London/Toronto/Sydney/1980

Library of Congress Cataloging in Publication Data

Fields, Alice.
    Insects.

      (An Easy-read fact book)
      Includes index.
    SUMMARY: Explains basic facts about insects including their stages of development, habits, and unique characteristics.
      1. Insects — Juvenile literature. [1. Insects]
I. Hurrell, David. II. Title.
QL467.2.F53   595.7   79-28095
ISBN 0-531-03244-2

R. L. 3.0 Spache Revised Formula

All rights reserved
Printed in the United Kingdom
6 5 4 3 2 1

Insects live nearly all over the world. During the warm summer months they are all around us.

Butterflies and **bees** are busy gathering **nectar** from flowers.

Grasshoppers leap out of the way as we walk in a quiet field.

Some kinds of insects get into our houses, too.

Flies are pests in the house on a hot summer day. They alight on the bread or fall into the milk.

Some insects spend all of their lives inside buildings. Insects such as **clothes moths**, **cockroaches**, and little **silverfish** often live in the house. You may see them running over the walls or floors in the kitchen or bathroom.

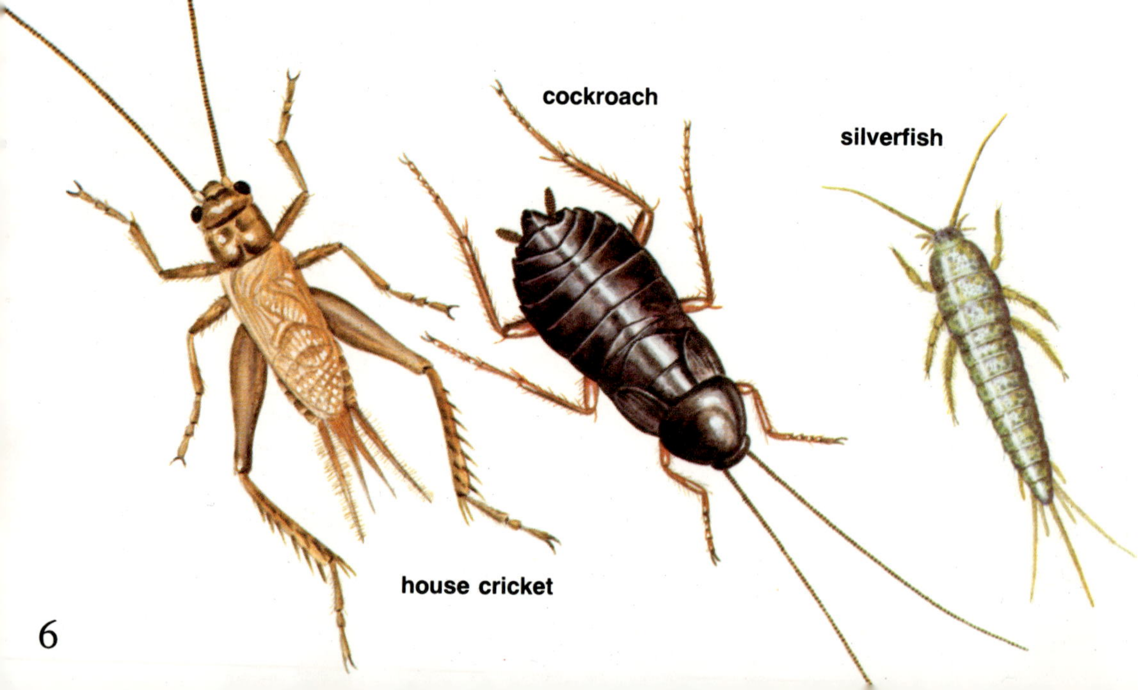

cockroach

silverfish

house cricket

Outside, insects are found nearly everywhere. They live on flowers and trees and under tree bark. They live in the soil and in ponds and lakes. They can even live on or inside other animals.

This picture shows just a few of these insects.

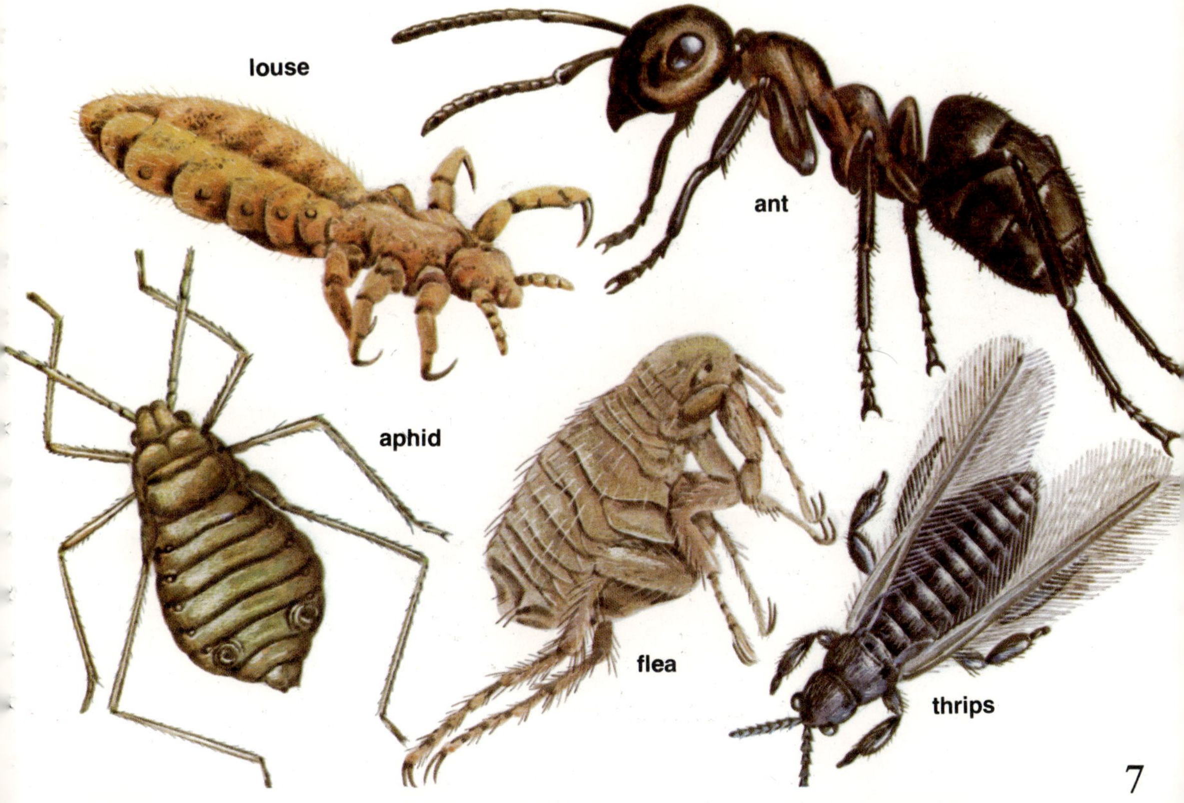

louse

ant

aphid

flea

thrips

Of all animals, insects are the greatest in number.
There are about 10,000 **species** (kinds) of **birds**. And there are 5,000 **mammal** species. But there are over 1,000,000 known species of **insects**.

Many kinds of insects exist that have not even been discovered or named. This is partly because many insects are tiny and hard to see. Others may live in places where people do not usually go. No one knows exactly how many insect species there are.

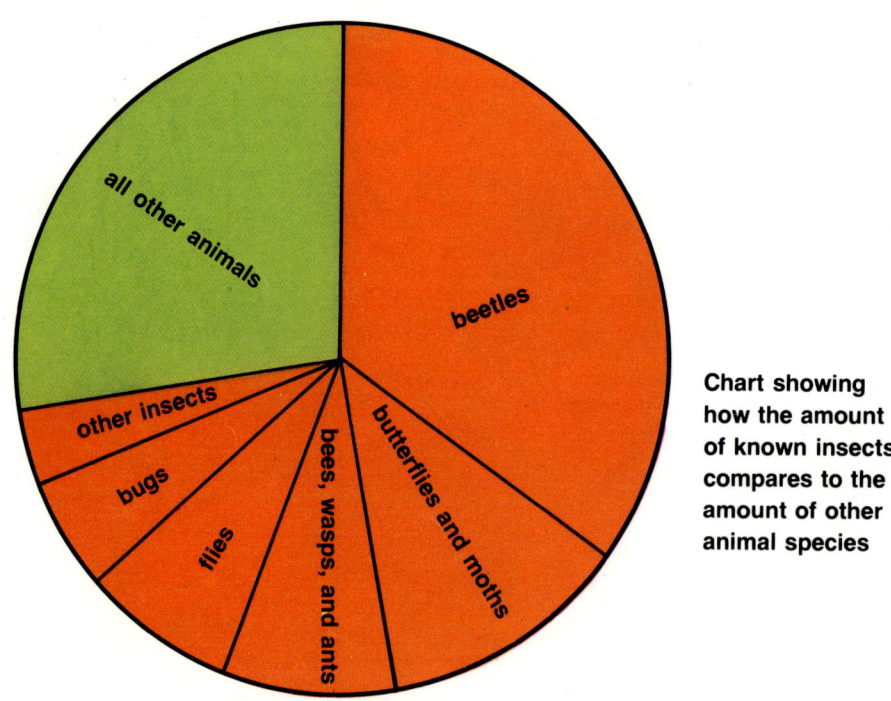

Chart showing how the amount of known insects compares to the amount of other animal species

A few of the many thousand kinds of beetles

Each species may contain thousands of different types of insects.

For example, there are about 300,000 different kinds of **beetles**.

Many kinds of small crawling creatures are mistaken for insects. At first sight they all look like "bugs."
Spiders, **pillbugs**, and **centipedes** are often called insects. But, in fact, they are not really true insects.

Just what is an insect, and how is it different from other animals? An **adult insect** has six legs.

But spiders have eight legs. Pillbugs have seven pairs of legs and centipedes have many more. Even snails are sometimes mistaken for insects, and they have no legs.

All these creatures belong to other orders (groups) of animals. They are not insects.

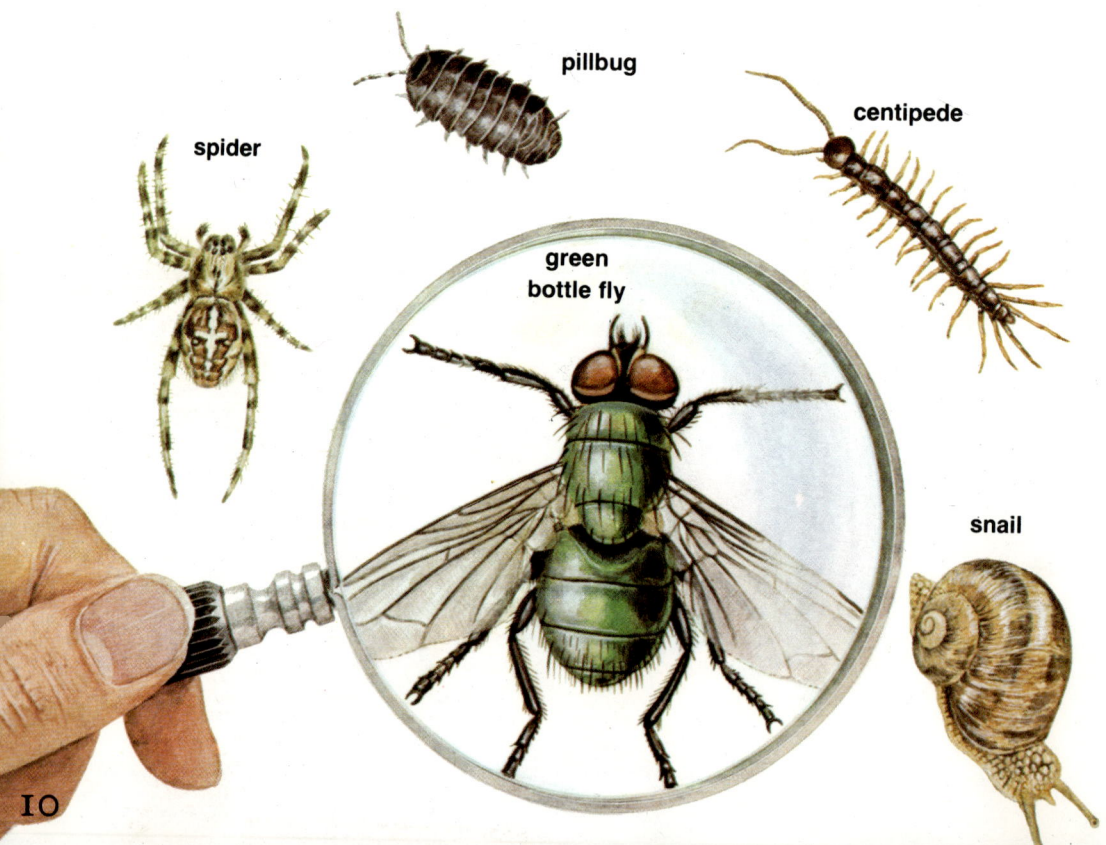

hoverfly

Most adult insects also have wings.

Flies have one pair of wings. Others, such as butterflies and **moths**, have two pairs.

Some insects have a wonderful talent for acrobatic flying. **Dragonflies** can dart and glide, and even fly backwards. Sometimes they reach speeds of 18 miles an hour (30 kph).

hawk mothwing

monarch, or milkweed butterfly

winged beetle

Another special thing about insects is the way they grow. They do not just get bigger like cats, dogs, and people.

Most insects pass through several stages. At each stage, the insect looks completely different from the way it looked before. These stages are known as **metamorphosis** (meta-MORE-fo-sis), which means "change of shape."

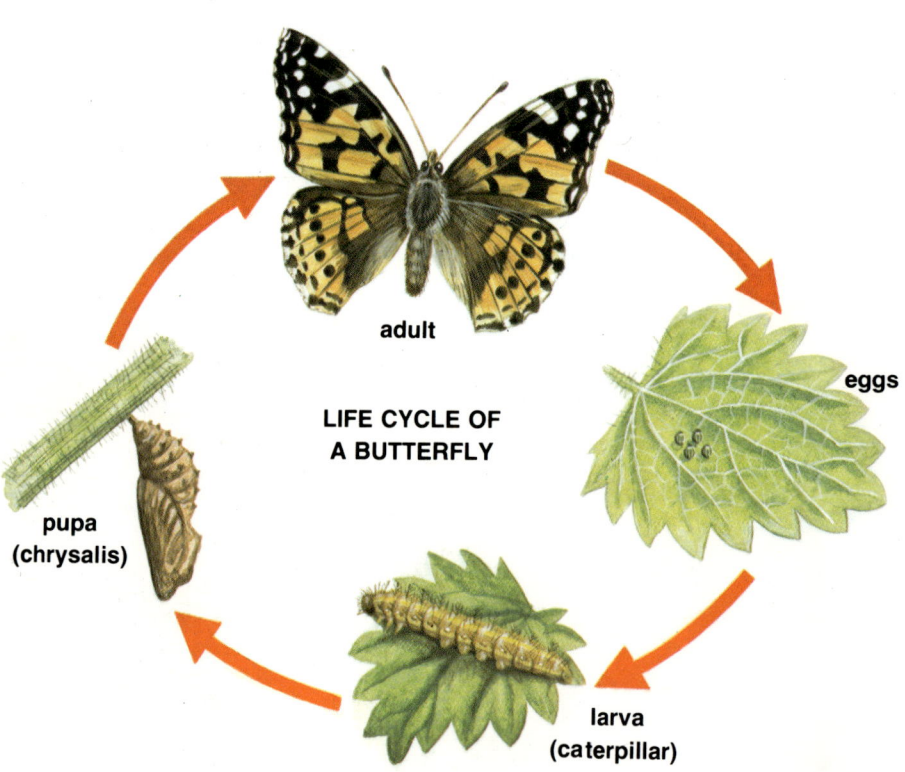

LIFE CYCLE OF A BUTTERFLY

Insect eggs
(very enlarged)

The insect **life cycle** begins with an egg.

Insect eggs are laid in all kinds of places—on leaves, in the soil, and in water. They are laid in animal fur and in birds' feathers. They are laid inside the stems of plants. Eggs are even laid inside the young or **larvae** of other insects.

In the next stage, these eggs hatch into larvae. All that an insect larva does is eat and grow.

Wasp laying eggs

Tunneling by larvae of elm bark beetle

As a larva grows bigger, it keeps shedding its skin.
     The larva is able to pump air under its skin. When it does this, the old, tight skin breaks open. Then, the larva pushes its way out.
     The new, looser skin lets the larva grow some more. At first, the new skin is soft and stretchy. Then, slowly, it gets harder and harder.

sawfly larvae

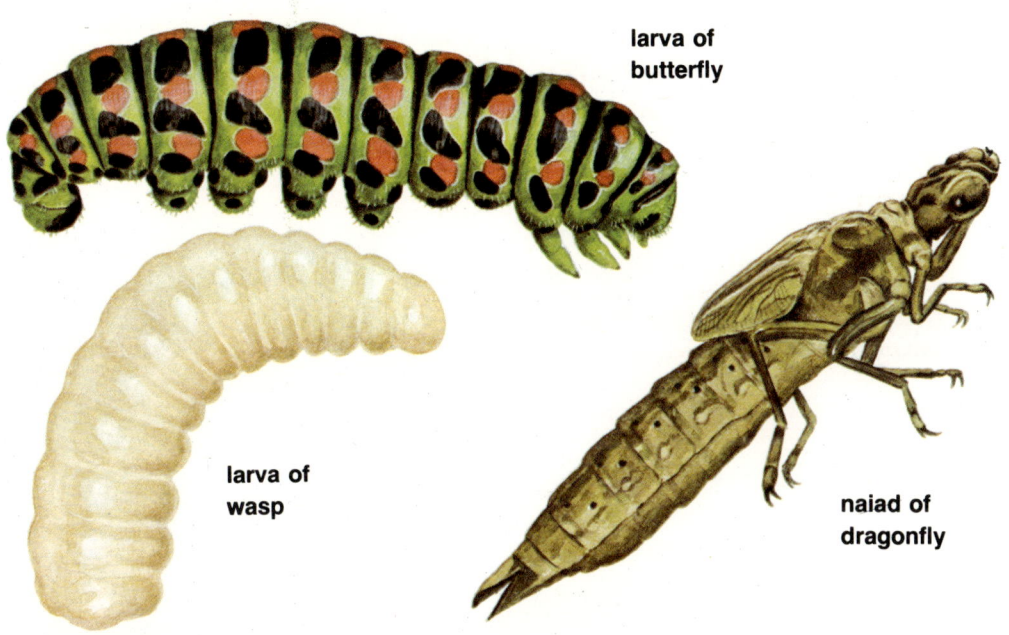

larva of butterfly

larva of wasp

naiad of dragonfly

Larvae may have more legs or fewer legs than they will have when they become adult insects.

Caterpillars are the larvae of butterflies and moths. They have many more legs than the adults. The extra legs of larvae are called "**false legs**."

In other insect groups, the larvae may have fewer legs. Or, they may even have no legs at all.

However, most insect larvae have the usual six legs.

The larva keeps on eating, growing, and changing its skin. When it becomes full-sized, it stops eating.

Among some groups of insects, a third change now takes place. The larva changes into a **pupa** (PYOO-pa) or **chrysalis** (KRIS-alis).

Butterflies, moths, ants, bees, and **wasps** are among the insects that go through this third change.

pupa of a wasp

chrysalis of a butterfly

section of a bee's nest showing an egg going through larval stage to pupa

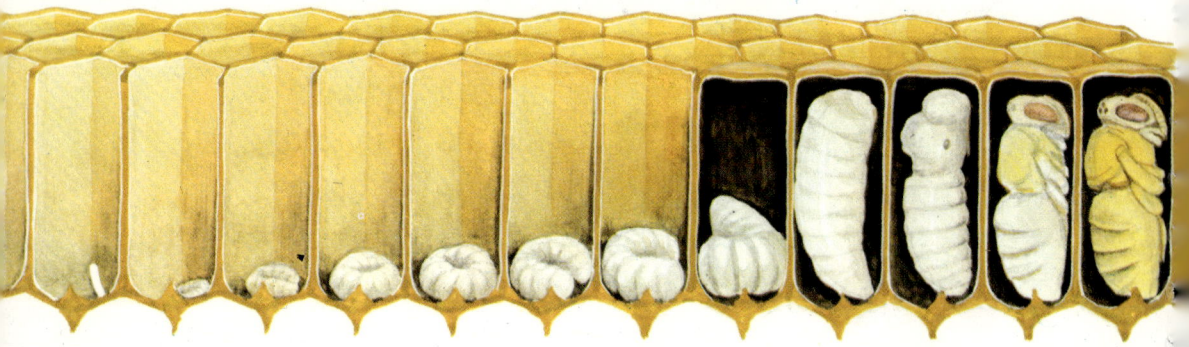

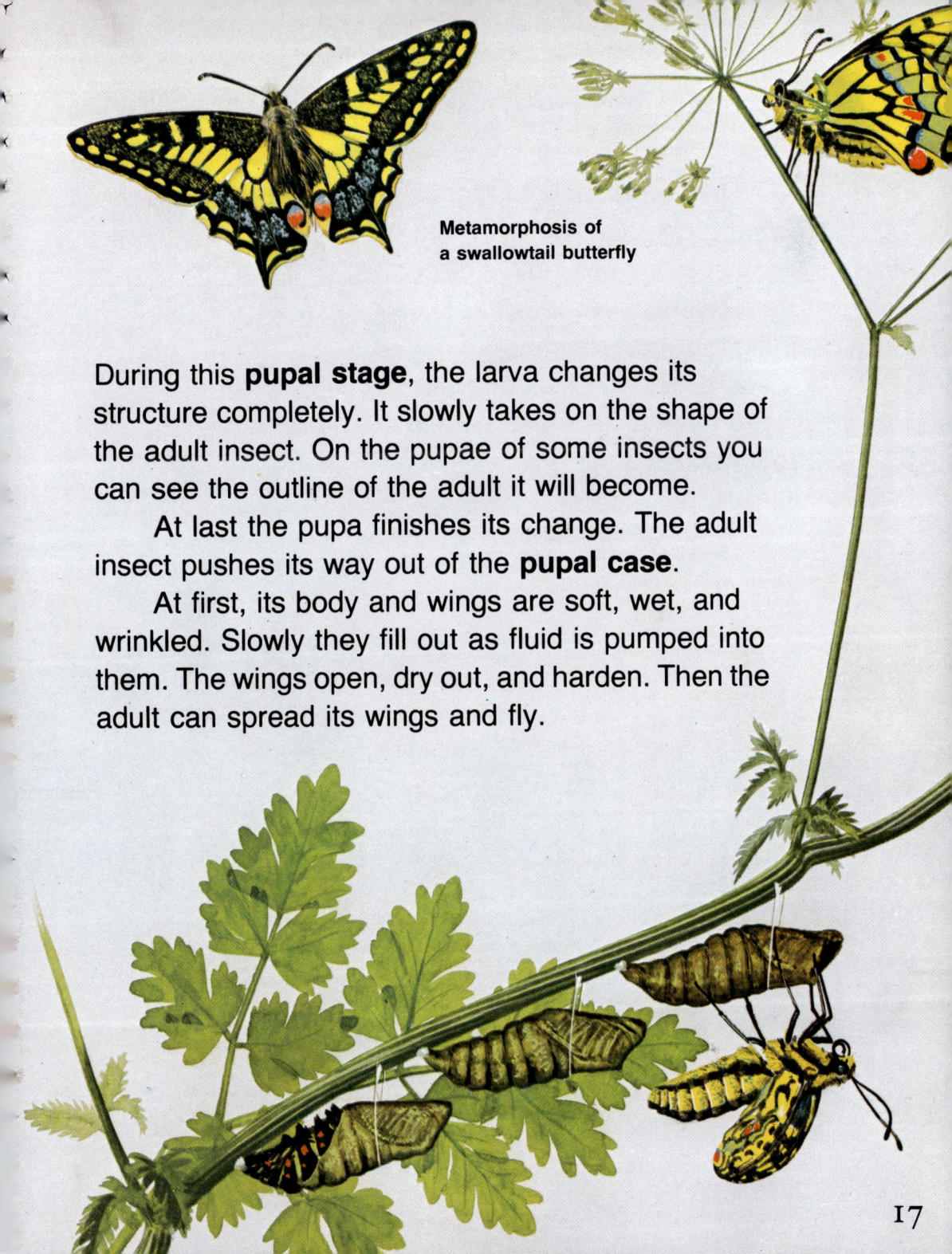

Metamorphosis of a swallowtail butterfly

During this **pupal stage**, the larva changes its structure completely. It slowly takes on the shape of the adult insect. On the pupae of some insects you can see the outline of the adult it will become.

At last the pupa finishes its change. The adult insect pushes its way out of the **pupal case**.

At first, its body and wings are soft, wet, and wrinkled. Slowly they fill out as fluid is pumped into them. The wings open, dry out, and harden. Then the adult can spread its wings and fly.

In other groups of insects there is no larval or pupal stage.

The young insect hatches from the egg and is called a nymph. The nymph grows by shedding its skin several times until it becomes an adult insect. Such insects include crickets, praying mantids, plant bugs, and cockroaches.

The nymph of a dragonfly is called a **naiad** (NY-ad) because it has gills to help it breathe underwater.

dragonfly coming out of naiad

two adult dragonflies— the female is laying eggs

dragonfly naiad

Adult insects exist to mate and lay eggs.
Although they may eat, they do not grow. A small fly does not grow into a larger fly, for example.

Many insects live for only a short time. The **mayfly** lives for only a few hours—just long enough to mate, lay eggs, and die.

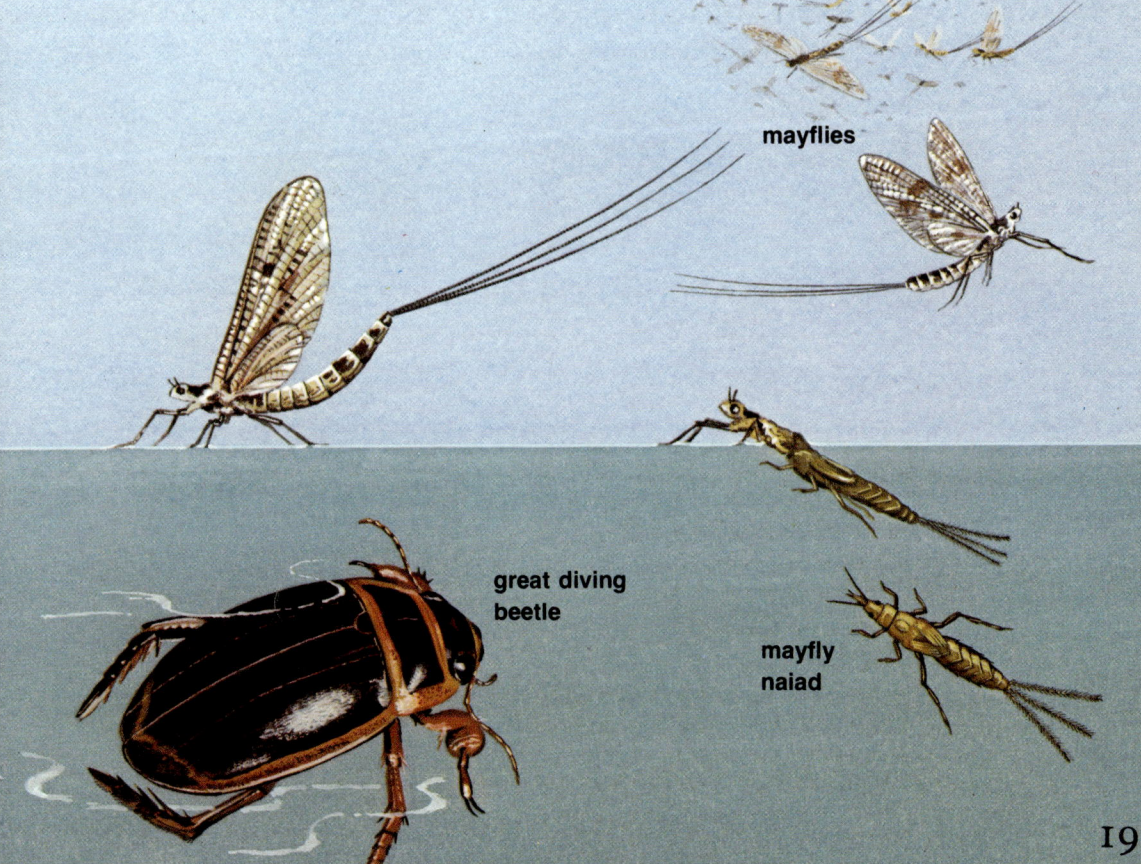

mayflies

great diving beetle

mayfly naiad

An aardvark attacking the thick hard walls of a termite's nest

Most insects do not take care of their eggs or larvae. But their young do not really need any help.

Some groups, however, build nests that hold thousands of insects. These nests are **communities** in which different jobs are shared. Some members look after all the young.

Insects that live in such groups are called **social insects** and include ants, bees, wasps, and termites.

Honeybees, for example, are social insects. Their nest or **hive** contains three classes of bees.

In the beehive there is one **queen**, some males (called **drones**), and many **workers**.

    The queen's job is to lay eggs.

    The drones are there to mate with the queen.

    The workers are females that are not able to lay eggs. Instead, they do all the work of the community. They gather **pollen** and **nectar** and make honey. And they build the hive and care for the young.

Bees storing nectar in the honey cells

A few **non-social insects** also look after their young.
   Certain species protect and feed the young nymphs. They stay close by until the young can take care of themselves.
   Earwigs, for example, act almost like hens with baby chicks!

Earwig protecting her eggs

Shield bug and her brood

Insect remains preserved in a piece of amber

Insects were on the earth millions of years before people.

Fossil (preserved) insects have been found in **amber** (hard tree sap) and in rocks. The fossils show us that some insect species have not changed. They still look the way they did long ago in **prehistoric** times.

Other insects from prehistoric times have changed or have become extinct (died out). In those times, for example, a huge dragonfly existed. Its wings were over 27 inches (70 cm) from tip to tip.

Today, the biggest insects are beetles. The great **African Goliath beetle** is about 4 inches (10 cm) long.

There are tropical **stick insects** over 11 inches (30 cm) long. And there is a moth with wings that stretch over 11 inches (30 cm).

African Goliath beetle, drawn to true size

Leaf miners leave a trail

Spangle galls

Plant galls (growths) caused by insect larvae

However, very large insects are unusual.

Most insects are small. Many are only a fraction of an inch (or a few millimeters) long.

Their small size lets insects live almost anywhere. Unlike larger animals, they can live in very small places.

Insects have a special way of breathing.

In mammals, **oxygen** is carried in the blood. But in insects, air enters through little holes called **spiracles** (SPEAR-akals).

The spiracles are on the sides of the insect's body. Each one opens into a small tube called a **trachea** (TRAY-key-a).

Each trachea has many branches which take the oxygen to all the body **cells**. You can see the spiracles very well on a large short-haired caterpillar.

The air moves slowly along the trachea. It spreads out through the body very gradually. This is a slow way of breathing.

If the insect is large, it takes a long time for oxygen to reach all parts of its body. So, large, fat insects cannot move very quickly.

Right: diagram showing the trachea (drawn in blue)

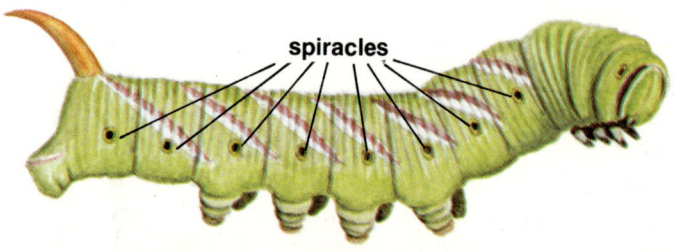

Insects are also slowed down by cold weather. They become **torpid** (slow-moving) and cannot run or fly. That is why we see many more insects during the summer. It is also why there are many more insect species in warm climates.

The eyes of insects are very different from ours. They are called **compound eyes**.

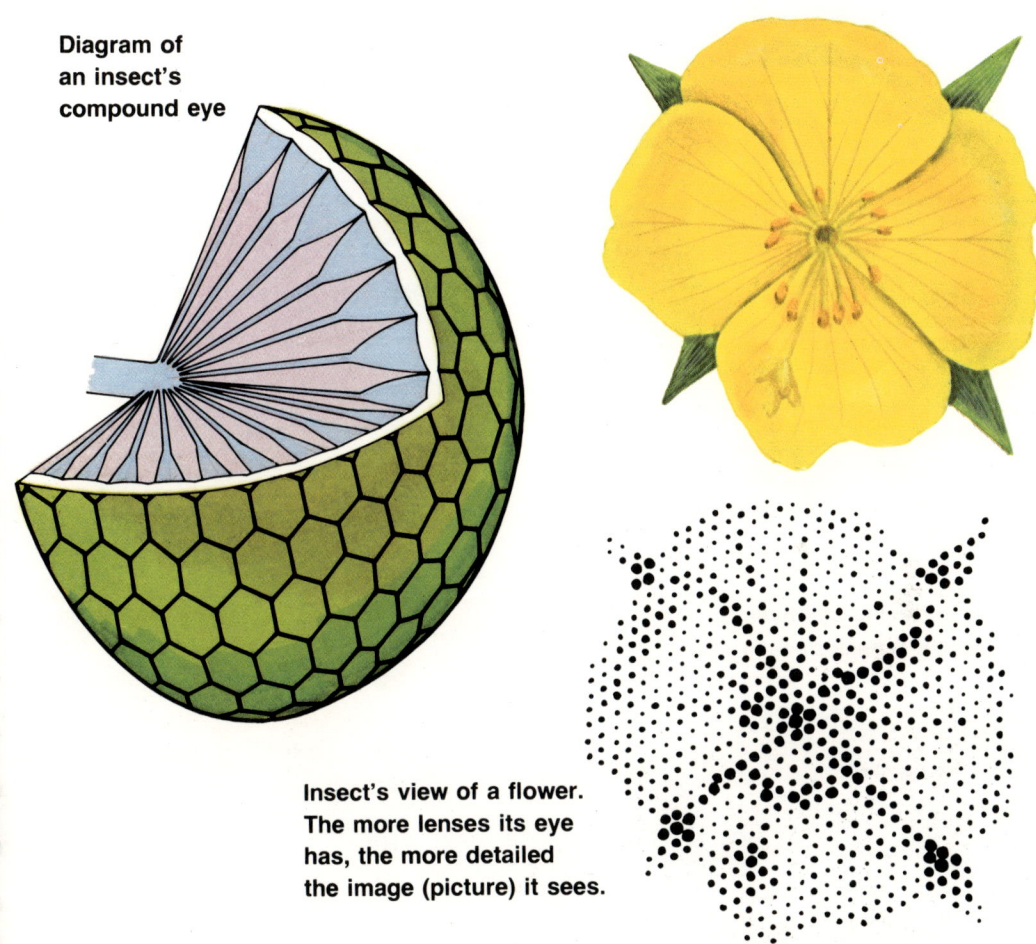

Diagram of an insect's compound eye

Insect's view of a flower. The more lenses its eye has, the more detailed the image (picture) it sees.

Compound eyes are made up of many little **lenses** joined together.

With each lens, the insect sees different parts of whatever it is looking at.

Our eyes can focus to see near or faraway objects. Insect eyes cannot focus as well.

But insect eyes are usually large and bulging. They can see the smallest movement very quickly. That is why it is hard to get close to insects like dragonflies.

With large eyes that stick out, the dragonfly can see almost all around it. Any movement causes it to dart away.

A dragonfly has over 20,000 lenses in each eye.

Single-celled eyes of a wasp

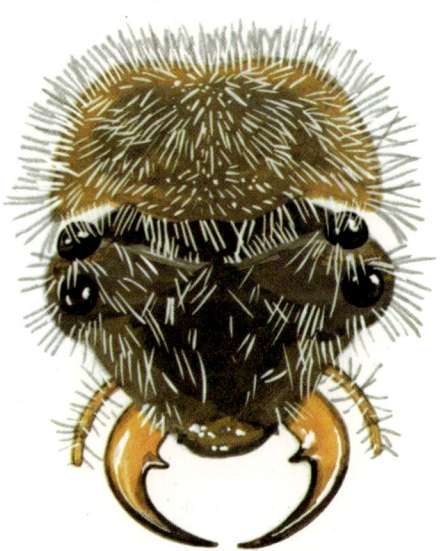

Single-celled eyes of a tiger beetle larva

All adult insects have compound eyes. But they may have other kinds of eyes as well.

These "extra" eyes have only one lens or cell. They are often set in a triangle between the compound eyes.

Unlike adult insects, larvae have only **single-celled eyes**.

Some insects see colors we cannot see. Bees and butterflies can see colors that show up only in **ultraviolet light** (beyond normal sight). Humans cannot see this kind of light.

Many flowers with nectar have colors in the ultraviolet range. These colors are like signal lights to bees and butterflies. They guide the insects to the nectar-bearing flowers.

Insects touch and smell mostly with their **antennae** (an-TEN-i). These are pairs of feelers on the head.

Male and female emperor moths

One ant can recognize another by feeling with its antennae.

A male moth, such as the **European Emperor**, usually has better antennae than the female.

The male uses its feathery antennae to pick up the smell of the female, sometimes from a great distance.

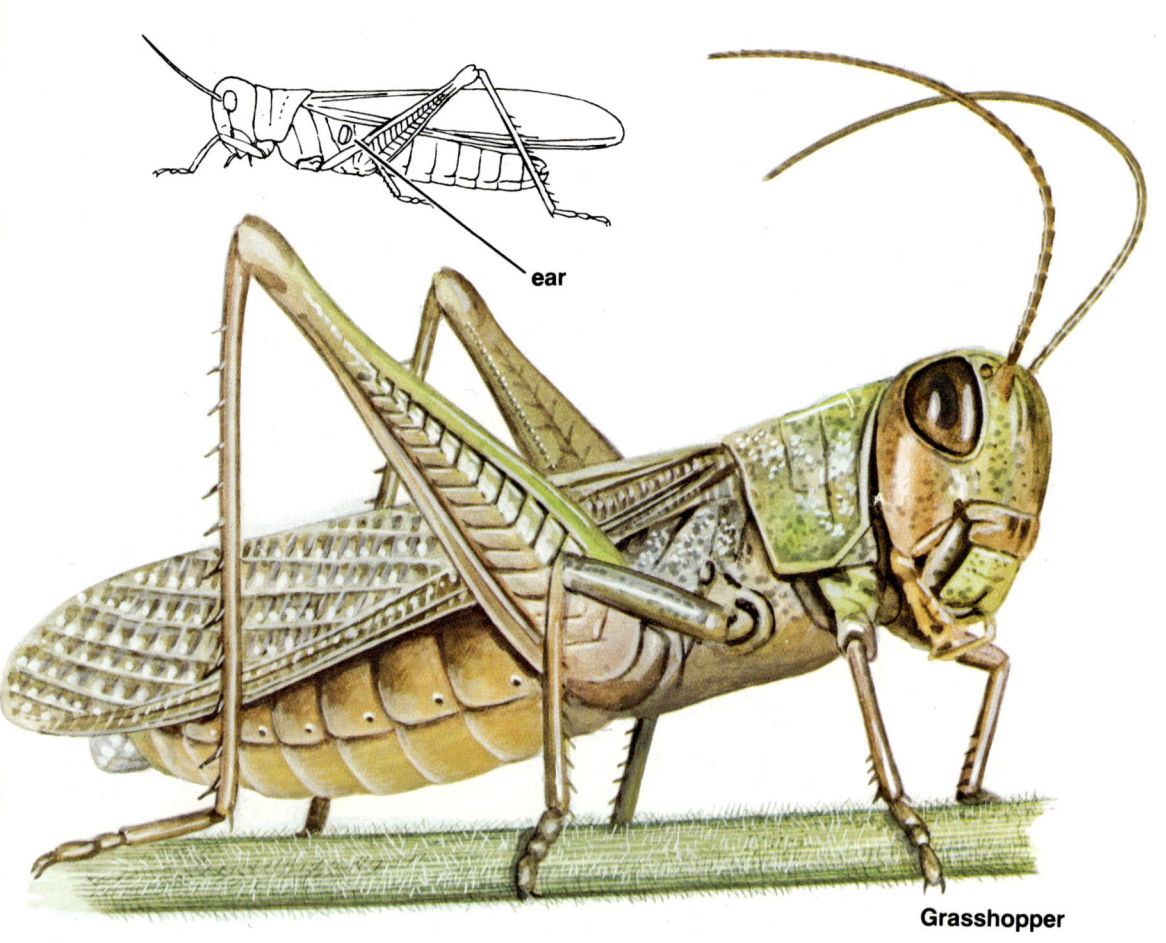

Grasshopper

Insects' ears are not like our ears. For one thing, they are not on the insect's head.

The ears of crickets are on their front legs. The ears of grasshoppers and moths are on the sides of their bodies.

Some insects make sounds to attract a mate.

Grasshoppers make shrill sounds. They do it by rubbing their back legs against their wings.

Crickets make chirping sounds by rubbing their wings together.

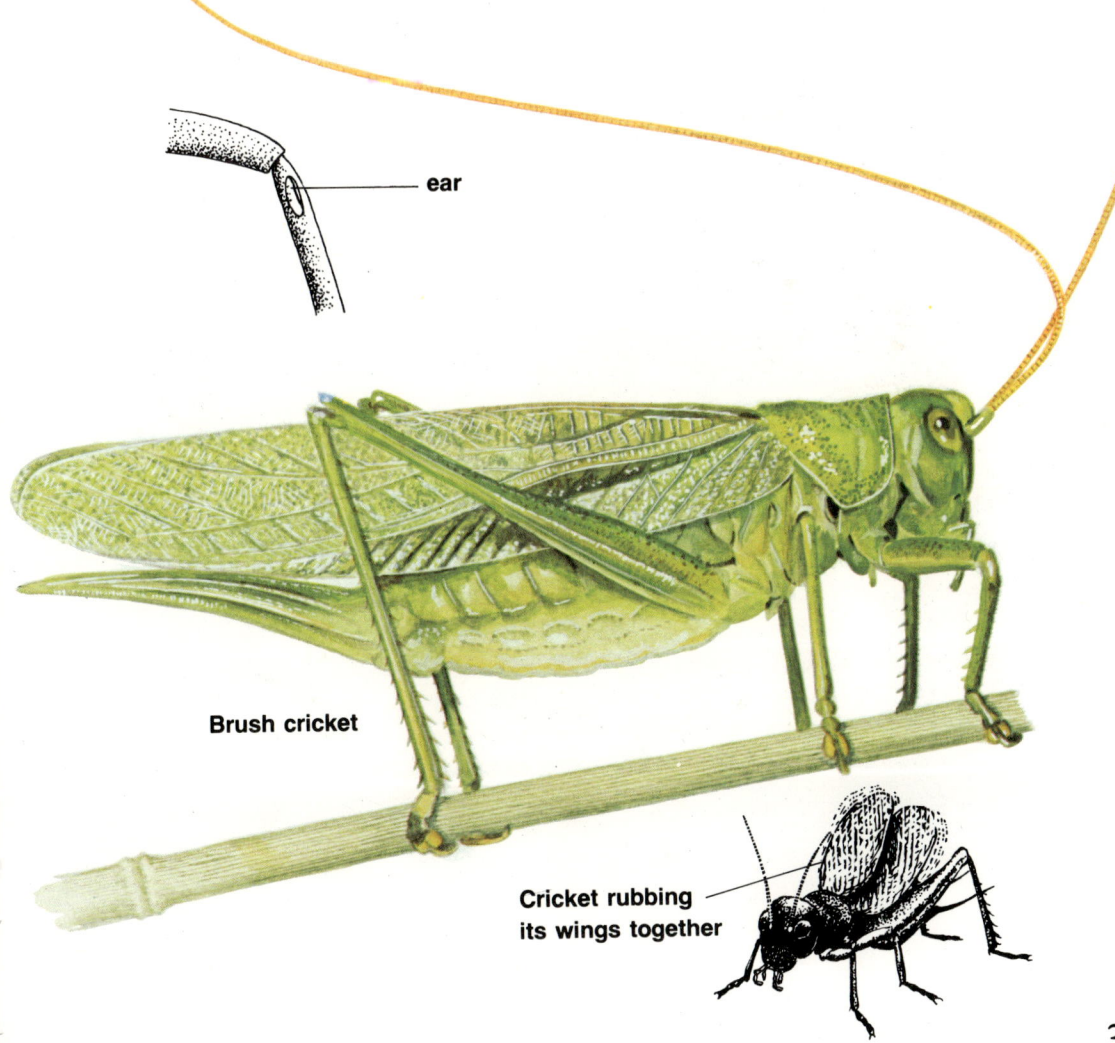

ear

Brush cricket

Cricket rubbing its wings together

Insects eat all kinds of food. They may eat leaves, flowers, bark, wood, nectar, pollen, or seeds. Some eat **lichen** or **fungi**, mushrooms, fruit, sap, blood, dead animals, or other insects.

We can often tell what an insect eats just by looking at the shape of its mouth.

The white admiral butterfly feeds on the nectar of flowers

Right: its proboscis (tongue) curls up when not in use

Tiger beetle eating a horsefly

Mosquitoes and **horseflies** have a long tongue called a **proboscis** (pro-BAH-sis). They use it to suck blood or tissues from other animals and people.

    Carnivorous (meat-eating) insects have strong jaws and legs. These help them to hang onto prey. Their jaws are shaped for tearing and biting.

    Among the carnivorous insects are **mantids** and many beetles, such as **tiger beetles**.

Some carnivorous insects eat dead animals that they find. This helps keep forests and fields clean.

Insects that eat plants have mouths shaped for eating plant fibers.

Grasshoppers and caterpillars are plant-eaters.

Sexton beetles burying their food

DIAGRAM SHOWING STING OF A HONEY BEE

poison glands

poison sac

sting

Insects, in turn, are often eaten by other animals. However, insects have different ways to protect themselves.

Ants, bees, and wasps can sting.

Grasshoppers and fleas can jump quickly and get away.

Others can fly away from their enemies.

Another kind of protection is **camouflage** (KAM-a-flaj). Camouflage is a way of using shape, and color, to hide something.

Insects often use their shape or color to blend into the background. Their enemies pass right by without seeing them.

Stick insects and **leaf insects** look like twigs and leaves.

stick insects

leaf insect

Above: this moth is camouflaged as bark. Below: the darker variety looks just like a blackened tree trunk.

Looper moth disguised as a branch

Caterpillars of many **looper moths** also look like twigs. They can hold onto a plant stem using just their **tail-claspers**. The rest of the body sticks out from the stem. It looks just like a small brown branch.

The wings of many moths look like tree bark. Others are beautifully marked like the veins of a leaf.

Sometimes color is used to frighten away or warn an enemy.

Some moths can suddenly flash brightly colored back wings. Others have markings that look like large staring eyes. These patterns may startle an enemy so the moth has time to get away.

Ladybugs have an unpleasant taste to birds. Their bright color helps to warn the birds. It is like a signal saying "This tastes terrible." Otherwise, the ladybug might get eaten by mistake.

When disturbed, the puss moth caterpillar leaves its camouflage, raises its head and body and waves its tail. It can also give off a strong acid fluid.

The North American Io moth shows its back wings when disturbed. This creates a startling effect.

Above: the hoverfly looks like a bee.
Right: the hornet moth looks like a wasp.

Instead of hiding, some insects fool their enemies. They imitate (act or look like) other insects.

A tasty insect may imitate one that tastes bad. Insects that have no stings may imitate those that do.

For example, some harmless beetles and flies look like wasps and bees. The enemy is fooled and leaves them alone.

Swarming locusts

People are among insects' greatest enemies because insects are **pests**. Some are very harmful to human beings. We often use chemicals to destroy these insects.

Mosquitoes can spread terrible diseases. Their bites can cause **malaria** and **yellow fever**.

Locusts, **caterpillars**, **cutworms**, and **wireworms** eat our crops. Gardeners and farmers must destroy them or lose their crops.

On the other hand, many insects are useful to us.
Bees make honey and **beeswax** (which is made into polish). **Silkworms** give us silk. And, far more important, bees and other insects carry pollen from one plant to another. This exchange of pollen makes new plants grow.

Other insects eat garden pests that kill our plants.
Lacewings and ladybugs eat sap-sucking **aphids**. Certain wasps and flies are natural enemies of caterpillars.

Other insects eat and break down waste matter. This helps to make the soil rich and fertile.

Insects are important members of the animal kingdom.

Without them, many trees and flowers would no longer exist.

Animals that depend on them for food would disappear.

Much good soil would become hard and unusable for growing crops.

The more we know about insects, the more interest, use, and beauty we see in them.

INDEX

Admiral butterflies, 36
African Goliath beetles, 24
Ants, 7, 8, 16, 20, 39
Aphids, 7, 46

Bees, 5, 8, 16, 20–21, 32, 39, 43, 45
    roles in the hive, 21
Beetles, 8, 9, 43
Bush crickets, 34
Butterflies, 5, 8, 11, 12, 16, 32, 36

Caterpillars, 15, 26, 38, 44, 46
Centipedes, 9, 10
Clothes moths, 6
Cockroaches, 6, 18
Crickets, 18, 34, 35
Cutworms, 44

Dragonflies, 11, 18, 24, 30

Earwigs, 22
Elm bark beetles, 13
Emperor moths, European, 33

Fleas, 7, 39
Flies, 6, 8, 11, 43, 41

Grasshoppers, 5, 34, 35, 38, 39
Great diving beetles, 19
Green bottle flies, 10

Hawk mothwings, 11
Hornet moths, 43
Horseflies, 37
Hoverflies, 11, 43

Insects
    breathing, 26
    care of young, 20–22
    defenses, 39–43
    description, 8–11, 17
    diet, 5, 21, 32, 36–38
    eggs, 13
    man and, 6, 44–45
    sight, 29–32
Io moths, North American, 42

Lacewings, 46
Ladybugs, 42, 46
Larvae, 12–16, 25, 31
Leaf insects, 40
Leaf miners, 25
Locusts, 44
Looper moths, 41

Mayflies, 19
Metamorphosis, 12–19
Monarch butterflies, 11
Mosquitoes, 37, 44
Moths, 8, 11, 16, 24, 33, 34, 41–42

Naiad, 15, 18
Nymphs, 18, 19, 22

Pillbugs, 7, 9, 10
Plant bugs, 18
Praying mantids, 18
Pupa, 12, 16–17
Puss moths, 42

Sawflies, 14
Sexton beetles, 38
Shield bugs, 22
Silkworms, 45
Silverfish, 6
Snails, 10
Spiders, 9, 10
Stick insects, 24, 40
Swallowtail butterflies, 17

Termites, 20
Thrips, 7
Tiger beetles, 31, 37

Wasps, 8, 13, 16, 20, 31, 39, 43, 46
Winged beetles, 11
Wireworms, 44